to **BE**

SAFE,

YOU should

ASSess

your

safety culture

(a workplace safety culture assessment guide)

by Thomas E. Williams, CRPO, CSS

illustrated by Justin T. Williams

to BE SAFE YOU should ASSess your safety culture

(a workplace safety culture assessment guide)

Written by Thomas E. Williams

Illustrated by Justin T. Williams

Disclaimer:

The information contained in this book was compiled from sources believed to be reliable and to represent the best current opinion on the subject. This information may be based on true and/or fictional events including, but not limited to, life experience, opinion, known and unknown facts, gossip, rumors, and tall tales. Some information may be exaggerated to stress importance or make a point. Names and titles have been changed to protect the innocent and the guilty. Any resemblance to real persons, living or dead or anywhere in between, is to be overlooked and not taken seriously. This information is intended to be used for informational and/or entertainment purposes. It is not written by a safety professional and therefore is not to be considered a gospel according to any known or unknown saint of any safety culture of any kind anywhere. An ergonomically correct environment, sufficient lighting, and neutral posture are recommended while reading this book. This book should not be read while driving or operating machinery including, but not limited to, heavy machinery and Smart cars. Contents may settle during reading. No warranty is expressed or implied. Not responsible for direct, indirect, incidental or consequential damages resulting from any defect, error or failure to perform. Void where prohibited. Use as directed. Apply to any affected area as needed. If condition persists, consult your supervisor. May be too intense for some readers. Freshest if consumed as soon as possible. Keep cool; process promptly. Subject to change without notice. All times approximate. Simulated pictures. Printing may, in time, fade and may disappear if erased. Pages may, in time, yellow and/or tear and eventually disintegrate. Keep away from fire or flame. Soggy when wet. Post office will not deliver without postage. No animals were harmed in the writing of this book. Beware of dogs. Reproduction strictly prohibited without the expressed written consent of the author. Any trademark mentioned herein appears for purposes of identification only. No assembly required. Batteries not included and should, in most cases, not be needed. One size fits all.

Contents

Contents

Preface

"Hey, can I go home and change shoes?"

This is the question I heard a maintenance technician ask the plant manager one morning. He explained the reason for his request by adding, "I didn't think I'd have to work hard today and I wore the wrong shoes."

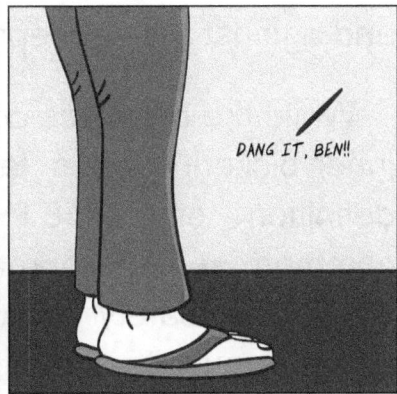

Just as my former coworker came to work that morning physically unprepared for the tasks at hand, many workers arrive every day unprepared to perform their job safely due to lack of training, lack of experience, and/or lack of concern on the part of management, coworkers, and/or the individual. The workplace safety culture has a major influence on how "safe" the worker is.

The term "safety culture" was first used in 1988 in a report issued about the 1986 disaster that occurred at the Chernobyl Nuclear Power Plant in Ukraine (major screw up – first ever "level 7 event" – involved over 500,000 workers). Since then, there has been a lot of research done about safety culture resulting in various definitions of it and even arguments for and against the concept.

While the brainiacs continue their studies and inner-bickering, let's take a look at a simple definition of SAFETY CULTURE, and a common sense approach to assessing your workplace safety culture that will hopefully provide you with useful information you can use to work more safely and develop a safer work environment.

(By the way...those letters after my name on the title page mean very little and are added for humor value. Seems that if you have letters after your name, you must really know what you're talking about and people are more likely to take interest in what you have to say. In my case, "CRPO" means Certified Refinery Process Operator (Chevron 1985) and "CSS" means Certified Safety Specialist (Safety Center, Inc. 2003). Neither is well known nor widely accepted. (I thought about adding "MD" for Mentally Disturbed, but I have no certificate to prove that one... yet.)

Introduction

Safety culture? What's that? First time I heard of it, I had to look it up. Sure, I know what safety is. After working nearly 30 years in industrial plants, I'd better know that. But what's this word, "culture" there for?

SAFETY CULTURE

Isn't a culture something the doctor takes to see if you've got strep throat or some other disease? Or, isn't culture stuff like expensive artwork, operas, fine wines, high-priced fancy restaurants, and other things that all those rich, uppity people like Thurston Howell, III are into? Doesn't culture also have to do with a common way of life shared by some ethnic, social, or age group? Like Roman culture, hippie culture, or youth culture.

Maybe that's it!! Could "safety culture" be some kind of way of life?

Well, a favorite source of info for this college dropout, *Wikipedia*, told me (in written word, of course), "Safety culture is a term used to describe the way in which safety is managed in the workplace, and often reflects 'the attitudes, beliefs, perceptions and values that employees share in relation to safety' (Cox and Cox, 1991)."

If a safety culture reflects common elements shared by employees regarding safety, but also is a term used to describe how safety is managed in the workplace, then whose responsibility is it, management or the real workers? And, how is the safety culture of a workplace assessed?

My answer to the first question is...

Although OSHA requires the employer to provide a safe workplace, it's *every* employee's responsibility to do what they can to help create and maintain a safe workplace. Don't just rely on others when it's *your* livelihood and life at risk. Look out for #1 – YOU!

I hope to answer the second question by the time you reach the back cover of this book (but please take it one thoroughly read page at a time to give me enough time).

We have our definition of "safety culture", but what elements do you consider in order to assess your workplace safety culture? Those aforementioned brainiacs are probably debating this question as you read this. I'm sure there are many more than this limited mind can think of, but I'd like to focus on six that seem very important to me (six is about all I can come up with anyway).

In this age of acronyms, I know one would be very helpful to offer at this point, but really...do we need another acronym? I could come up with one, I'm sure, but it would probably be confused with the same acronym that means something entirely different.

For example – I once received a message from a good friend of mine. The message simply read, "BFF". I thought she was saying that we'd be best friends forever. But, what she wanted me to know was, "Burying Fred Friday"…I missed the funeral.

So, no acronym this time – vowels! Yes, that's right, your eyes do not deceive you, and your reading is not impaired! We will use the vowels of our American English language (including the "sometimes Y") to identify elements to consider when assessing safety culture. How easy is that?!! Simple minds create simple ways!

Unrelated story: One of those Thurston-types once called me a "simpleton". From his arrogant demeanor, I'm sure he was confident that he knew what he was talking about. But, I showed him. I told him, "That shows how little you know. I only weigh 200 pounds. It would take ten of me to be a simpleton." (And, 11.2 of me to be a long ton – I'm pretty good at math!)

Now, let's get to those vowels…

Chapter 1

"A" is for Attitude

For those of you who learned early in life that "A" is for apple, please don't let me confuse you. It is, but for this book, "A" is for attitude. Some will argue that "A" should be for awareness, but the *attitude* will make or break whatever level of awareness exists, IMHO (OMG an acronym – no, two!!) If you disagree and still insist that 'A' should be for awareness, write your own book.

Anyway, ATTITUDE, yeah, I have one, do you? And, what an appropriate place to start with our elements to consider for assessing a safety culture because...your *attitude* will affect how you perceive the importance of each element as well as safety in general.

For a good safety culture you of course need a positive attitude toward all safety-related matters. How's the attitude toward safety in your workplace? Do you have an attitude like, "We're safe enough – there's never been an accident here!"? Or maybe, "Accidents don't happen to me". Hey, those are very common ones, so stop kicking yourself because you share them, too, and just realized they're bad!

Those were also the attitudes I found common among management and a few coworkers at my workplace. For over twenty years the plant had not had a lost-time accident – so easy to become lax on safety!! Then early one fine spring morning a near-fatal accident happened at my plant on my shift...to me!

After experiencing the heat of a 5,000-volt motor starter arc-flash (flash suit not included), I've warmed up to an extremely positive attitude toward safety and highly recommend possession of same to everyone.

(Maybe if the management types at my workplace had felt the heat that I did, the accident would have fired up their attitude as well. But, sadly it only put them on a low simmer for a short while. And, I'm pretty sure that simmer was a result of the heat put on them by upper management. But, that's just my opinion based on my observations.)

Your safety attitude is your choice whether you're part of management or one of the real workers. And, if you're a management type, don't be surprised if your subordinates adopt the attitude they perceive you to have be it a positive or (hopefully not) a negative or ho-hum attitude. You are, after all, in the limelight and setting an example and "mood" for your workforce.

THE PLANT WAS IN AUTO... FIGURED I'D TAKE A NAP.

Recommendation: If you need an "attitude adjustment" regarding safety, ask yourself this question each time you go through that door, gate, turnstile, or whatever portal gets you into the workplace...

"Do I want to continue making memories with my loved ones? Or, do I want to BECOME A MEMORY to my loved ones?"

How's your attitude now?

Chapter 2

"E" is for Environment

So, what's air, water, those trees we love to hug, and that protected species of tiny polka dot-toed brine shrimp in your soil have to do with safety? Nothing that I know of. It's the *work* environment I want to focus on in this chapter, and since several other characteristics of the overall work environment are covered in other chapters of this book, just a look at a few physical conditions should do for this one. So, let's get an idea of how safe your workplace surroundings are...

When you arrive at work, what do you see?

Can you get from point "A" to point "B" without a slip, trip, or fall? Or, do you have to do long jumps and cartwheels to avoid things such as spills, open pits, uneven surfaces, obstructed walkways, or, in the office or control room, open desk drawers, open file cabinet drawers, improperly placed boxes, and blocked exits? Don't forget to watch out for hoses and cords as well. Hey, what's that live extension cord doing lying in a puddle of water?!! Shocking!!

Got hazardous stuff like flammables and chemicals?

Do you know what's on site, where each is located, and the hazardous properties of each one? Are all containers of flammables stored in a self-closing, FM approved flammables cabinet? Are there "Flammable" and "No Smoking" signs on the cabinet? Are tanks, drums, and all other containers holding hazardous materials clearly marked with its contents and are the proper HazMat labels also affixed (I think that means stuck on it somewhere)? Do you know what an MSDS (Material Safety Data Sheet) is? Is there one on site for each substance? Where are those MSDS's, anyway??

If your workplace has hazardous stuff in pipes running all over the place along with piping for various processes and other systems, how many different colors of pipe do you see?

If they're all the same color, there's a problem. If there are lines painted a separate color to designate a hazardous material, that's good. If that line is labeled with the name of the material, that's even better. But, if the colors of all the plant piping are many and have you looking for a pot of gold at the end, that's great!! The more color coded pipes you have and the more labeling you see indicates there's a good chance a good safety culture may exist in your workplace. (Or, maybe it just means your management has an artistic flair about them – you be the judge.) If direction of flow is also labeled, wow, somebody knows their stuff!!

ANSI standards for color coding pipe service are green for water, red for fire extinguishing fluids, orange for toxic and corrosive fluids, yellow for flammable fluids, brown for combustible fluids, blue for compressed air, and black, white, gray, or purple as designated by the plant. (The correct colors to use actually have "safety" preceding the color, e.g., safety green, safety red, etc.) Now, go forth and paint the plant!

Got an old rusty 98% sulfuric acid line running overhead that may burst and shower everyone around it before you've finished reading this page? Wouldn't a well-maintained line that runs at ground level, is painted orange, and labeled with "Sulfuric Acid – 98%" and direction of flow contribute to a much safer work environment?

Insulated lines are usually a sign of something pretty hot or something pretty cold and are insulated to keep them that way. Steam and other hot fluids should be insulated. Chilled water is often insulated. That insulation improves system efficiency, but also provides protection from the extreme temperature. Anything hot and within seven feet of the floor or working surface must be insulated according to OSHA.

Also, when looking around the workplace, do you see fire extinguishers, first aid kits, safety shower/eyewash stations, spill containment kits, or other emergency equipment? Is it all clearly marked, unobstructed, and easily accessible?

What about safety signage? Are hazardous areas identified with "Danger" or "Warning" signs? Are there also signs to tell you what the hazard is and what personal protective equipment (PPE) is required in the hazardous area? I'm not just talking hazardous materials here. A high voltage area, high noise area, any area containing a hazard of any kind is a hazardous area and should be identified.

Does your workplace incorporate engineering controls whenever possible to reduce hazards? I've already mentioned insulated lines. That's an example of an engineering control. Are high temperature surfaces insulated? High noise areas or extremely loud equipment should also be insulated whenever possible. Are there guards installed on machinery to protect you from hazards such as those created by point of operation, rotating parts, flying chips, and sparks? Is your office, shop, control room, or lab properly ventilated? Can a hazardous material be replaced with something less hazardous? Or, can the onsite quantity of a hazardous material be reduced? These are examples of engineering controls used in the workplace. If a hazard cannot be removed from the workplace, engineering controls should be the first choice to reduce exposure to the hazard.

Next in the line of defense against hazards are administrative controls. In your workplace, is access to highly hazardous areas restricted? Does all equipment appear to be maintained properly? Is there evidence of a good housekeeping program in place? Much of what we covered earlier in this chapter, safety signage, color-coding and labeling, and removing slip/trip/fall hazards, are all types of administrative controls.

(Work practices are often considered to be administrative controls as well. In all workplaces, safe work practices are extremely important. These are discussed in chapter 3 on implementation and chapter 5 on understanding.)

Where engineering controls and administrative controls fail to eliminate exposure to the hazard completely, is PPE made available for your use? Do you know how to use it and know its limitations? (OSHA requires the employer to provide necessary PPE and to train their employees in the use, care, and limitations of that PPE.)

Now that you've looked around your workplace, do you feel safe enough? Are hazards removed, controlled, or otherwise in check? Do you have the PPE you need to prevent personal injury? And, do you have the proper equipment and supplies to deal with emergency situations? Did I hear a "No" answer in there somewhere? If so, take action to improve your work environment. If you're a management type, make the necessary changes. The rest of us, if it can't be fixed on the spot, have to be content with reporting it to our supervisor hoping our concern will be taken seriously. In that case, document your concern via email, work request, or safety suggestion as proof that you did your part!

The physical characteristics contributing to a safe work environment are probably the easiest to get a handle on, but too often are neglected or overlooked entirely.

Chapter 3

"I" is for Implementation

Have policies and procedures been implemented in your workplace to comply with all relevant regulations (OSHA, NFPA, State, Local, Corporate, etc.)? Are new policies and procedures implemented in a timely fashion as they are received by local management? Are they communicated to the workforce properly and does management have documented proof that the communication or training has been received and understood by each affected employee?

If you're not up to date on compliance, management can expect a hefty fine to pay should you be audited or have an accident. It's management's responsibility to provide and implement what's needed related to safety. This includes safe work procedures and policies, engineering and administrative controls to reduce hazards, and any required

17

personal protective equipment (PPE) for employees.

Does management preach, "Safety First", but when it comes right down to it, is management too concerned with their "bottom line" to put safety first? Sure, management expects safety first of their employees, but do *they* practice it as well?

To the management types –

If the costs involved in creating and maintaining a truly safe workplace are a major concern, that is, if you're worried about its effect on your "bottom line", may I suggest that you practice the "line" (i.e. Safety First) and it will protect your "bottom" (i.e. CYA – if you're unsure of what that acronym means, ask one of your real workers and they will explain it to you; they've probably seen management do it many times and can offer several examples to help you better understand "CYA").

Ben realized much too late that, due to budget cuts, his team could only afford ten helmets.

Chapter 4

"O" is for Oversight

Oversight – now there's an interesting word! Check out its two definitions:

1) Supervision; watchful care or management
2) An omission, mistake, or error; especially one made as a result of a failure to notice something.

(My "metaphrased" definitions after having consulted several dictionaries.)

So, if I understand this word correctly, *lack* of oversight *can cause* an oversight? Wow! Imagine That!!

When assessing your workplace safety culture, consider how good your "definition #1" oversight is and how often "definition #2" oversights occur.

In the words of former president Richard Milhous Nixon, "Let me make this perfectly clear"…

Any oversight shows lack of oversight and proper oversight prevents oversights. Got it?

Need I say more? Probably not, but I will anyway (it's my book).

Before I get into oversight by management, let me point out that "oversight" (i.e. watchful care) is *extremely* important at the "individual" level as well. After all, the individual should at all times exercise watchful care over self and coworkers. (See Chapter 1 on attitude again if necessary.)

At the plant where I work, even though I'm among the lowest life form in our workplace food chain, I'm allowed to stop any activity by anyone in the plant if it's unsafe. This policy has been practiced at every plant I've worked at without exception. *I highly recommended that same authority be given to all employees by their management.*

However…

WARNING TO THE REAL WORKERS:

If you are given this authority by management, USE IT! If you fail to stop any unsafe activity you observe, it will come back to haunt you. You *will* be management's "scapegoat". "YOU should have said something and stopped it", they will say. You could have, and had the authority to, so why didn't you?

THE COMPANY ALWAYS DENIED IT, BUT BILL ALWAYS KNEW HE WAS THE SCAPEGOAT.

I've seen this scenario play out a few times in my lifetime, so beware! Scapegoats are in some safety cultures freely created by a management that is lacking in definition #1 oversight and unwilling to recognize and/or take responsibility for their own definition #2 oversight. (Yet according to management, scapegoats do not exist. To management they are a mythical creature like the satyr!)

Oversight (i.e. watchful care and supervision) by management should be done at all levels - corporate, regional, and local – as needed to insure as safe a workplace as humanly possible. Each level should at the least annually audit the safety performance of the next lower level.

Oversights (i.e. mistakes, errors, etc.) are a reflection of the degree of management's dedication to safety and cause unsafe conditions and accidents that far too often result in injury or death.

To sum it up…your oversight should prevent oversights. If and when it doesn't, correct as needed to achieve the intended purpose.

THE DEPARTMENT OF OVERSIGHT OVERSIGHTS

Chapter 5

"U" is for Understanding

A little story from my personal experience...

I worked for a manager who had been at the plant for over twenty years. He had been there even before the plant was commissioned. Before his promotion to management, he was a plant operator. A legend in his own mind, he could tell you everything about anything related to our plant.

I submitted a safety suggestion that a plant hazard assessment be done. No one at the plant could show me one nor had anyone seen one. It is required by OSHA to have one – a documented one. (No plant hazard assessment for a plant that had been operating over twenty years?!! What's that say about the workplace safety culture?)

Rather than put his twenty years plus of experience at that plant to use and create a hazard assessment, the project was turned over to a few plant operations and maintenance technicians who were members of the plant safety committee. That plant hazard assessment is management's responsibility, and while management can delegate anything to any subordinate, wouldn't it have made more sense for the manager to have just gone ahead and done it? Who would it seem is better experienced to do it? And, wouldn't he be reviewing it for accuracy anyway before it's issued? (Or would he?)

Did he not do it because he just didn't want to? Or, did he not understand its importance related to plant safety? Or, did he himself not understand the hazards in the plant he had worked in for twenty plus years and didn't want that embarrassing fact found out? Maybe I should give him the benefit of the doubt and suggest that he just wanted to see what his subordinates knew.

Understand the importance of safety and give it its proper priority – FIRST! This should be practiced by everyone in the workplace regardless of position.

Our understanding is only as good as the information we possess. Understanding safety may be defined as applied knowledge of regulations, procedures, and existing hazards gained through experience, research, and training. A good deal of common sense is very helpful as well.

Evidence of understanding can be seen in things such as the practice of safe work procedures, PPE usage, and workplace safety records. (Don't forget to include all those "near miss" occurrences.)

Are management's expectations regarding safety clearly communicated to everyone in the workplace? Does each employee know their individual responsibilities, know each hazard in the workplace, and do they know and understand all applicable safety procedures and policies? Have they been trained in emergency response for incidents such as spills, fire, or employee injury? Perhaps more importantly, does management know these things? If the management types don't understand, why do they expect the real workers to understand it all?

To promote "understanding", is a safety orientation given to everyone that comes to the workplace before they are allowed to begin their work and roam unescorted around the site? This should include new hires, visitors, and contractors. Make certain that *everyone* in the workplace understands the hazards and safety requirements.

To ensure "understanding", are written quizzes given to, or questions asked verbally of, employees at the end of a safety training session to make certain employees understand what's been covered and are sufficiently trained? Do you do the same following a safety orientation?

Or, do you wait until an accident happens and then realize the failure of the trainer to teach because the student hadn't learned?!!

Remember, you can park your truck at the gas station, but you can't make it pump the gas. Oh wait, that's supposed to be something about a horse and water, isn't it? Anyway, what I want to stress is…you can have all the talk and paperwork that it takes to "snow" the regulatory agencies, but if understanding is lacking when it comes to all things safety related in your workplace, especially the importance of safety, your safety culture is terribly lacking and is, in itself, a safety hazard!

Chapter 6

"Y" is for You

"You talking to me?" you may be asking. Yeah, absolutely, I'm talking to you! Management type or real worker, *you* are an important element in the workplace safety culture. This "Y" is not just included "sometimes"; it must be included *all the time!*

As I stated earlier in the chapter on implementation, it's management's responsibility to provide and implement what's needed related to safety. It's *your* responsibility to work safely by following safe work procedures, using required personal protective equipment (PPE), blah, blah, you get the idea. (Sometimes I can sound like a management type...)

Your awareness; your attitude; your knowledge and practice of safe work procedures and policies; your usage of required PPE; your recognizing, reporting, correcting, and/or isolating any unsafe condition; your looking out for the safety of others as well as yourself – there's an awful lot that *you*, the individual, can contribute to a positive safety culture in *your* workplace.

Let your example lead, not mislead. Don't shortcut safety. You never know who may be watching you. It may be that new employee who has a lot yet to learn, or it may be a management type hoping to catch someone in an unsafe act and make an example of them to "prove" that management's on top of things. (Did someone say, "Scapegoat"?)

Take safety home with you. Safety awareness is important away from the workplace as well. Help keep your family and friends safe by sharing what you know. Let your example lead there as well. Would you want a loved one to have an accident because you hadn't said something you could have that would have prevented it or because they watched your misleading example? It's comforting to know that your loved ones are safer because of your example and shared knowledge.

"Safety begins with you", I've heard it said. Kind of confuses me – looks to me like it starts with an "S". I do know "You're fired!" starts with "you". Maybe that's a good reason to be safe!

Conclusion

Management types, what's your approach to safety? Proactive or reactive? Take a "P", you'll feel better. Being proactive in safety related matters will help prevent accidents, increase safety awareness, and contribute immensely toward an exemplary workplace safety culture.

Got a reactive approach? Why? What's your problem? Can't be bothered by safety? It costs too much? Is it too complicated for you to understand? What is it that keeps you from being proactive? You're the type that waits to comply and implement after an accident, injury, or death - whatever it took to make you CYA. Create any scapegoats lately? You wouldn't admit it if you had! How can you sleep at night? Have you no conscience?

Your employees deserve better than you!! You're the kind that needs a nice, hefty, bank account-emptying fine. Maybe then you'd see the error of your ways before another injury or death happens. As you now are, you don't belong in the workplace. *You are just another hazard.*

Okay, okay, I'll settle back down, but let me also say this:

"Zero loss-time accidents" is a safety record that'll "make you proud", but it doesn't necessarily mean there's a good workplace safety culture. It could be a result of good luck coupled with too few opportunities for a serious accident to occur.

Your workplace safety culture may be excellent, it may itself be a safety hazard, or it may be somewhere in between. As you read this book, did you find yourself already assessing your workplace and yourself?

I hope the elements we've discussed and questions asked in this book will help you easily assess your workplace safety culture and provide ideas of how to make yours an exemplary one. And, don't forget your own personal "safety culture". The best place for an exemplary safety culture to start is with you – management type or real worker.

Back in the Introduction of this book, I defined "safety culture" using a definition from *Wikipedia*. As I was writing this book, I thought of another, probably easier to remember, definition of safety culture and it even contains an acronym! (See, I told you I could probably come up with one!!)

Here's *MY* definition of "safety culture": *The way in which safety is PIMP'd in the workplace.* Or, (speaking directly to the management types now), the way *you* PIMP safety in the workplace. What's PIMP? Well, it's an acronym commonly applied to someone managing, shall we say, a "working harem", but in this case:

PIMP = Perceive, Implement, Manage, and Practice.

And, what a fitting acronym it is! After all, isn't safety something that good management wants to "sell" to the workforce? (Hmmm... "How to PIMP Safety in the Workplace". Sounds like I will start on another book.)

Thanks for taking the time to read what I've had to say. Please give it all serious consideration and do what you can to improve the safety culture at your workplace if improvement is needed. It has been said, "There's always room for improvement."

If you've learned nothing new by reading this book, I don't apologize, but will say to you, "Thank you." We need more people like you at all levels of the workplace in many of the companies around the world.

Stay safe, and continue making memories with your loved ones!

Epilogue

Gone in a Flash

Early one morning as I was working as an operator technician at a cogeneration power plant the plant 5,000-volt (5kv) bus tripped. That 5kv bus had tripped many times in the previous week, but management decided that we would continue to run the plant as long as possible. This was the fourth trip in about a twenty-six hour period. We had been through the same upset condition the previous morning and it had happened twice to the night shift as well.

I went to the Motor Control Center (MCC) to reset breakers. After resetting two or three breakers, I was attempting to reset another one when it arc-flashed. I heard a loud "bang" and the breaker door blew open. Immediately, there was total blackness around me except for an orange ball of fire coming out from within the breaker cabinet. It all happened much faster than words can properly describe. Perhaps, "in an instant" best describes it.

(The plant lost all power and all equipment tripped. The arc-flash and resulting fire destroyed other 5kv equipment as well. The entire plant was down and offline for nearly three days while necessary repairs were made to allow production to resume. One manager told me the accident cost the company over a half million dollars in lost production alone.)

The heat of the arc-flash was intense and I could feel myself burning. All I could see was total darkness and thought I'd been blinded. Within a few seconds, I could see through the smoke and saw that the breaker was on fire. The fire extinguishing system released Halon into the room. I exited the MCC and went back to the control room. The on-shift supervisor called 911 from his personal cell phone since the power was out and the plant phones were not working. Another coworker brought me a spray and some packets of burn lotion that I applied to the burns. The EMT's arrived and took me to a hospital burn unit where I was treated and stayed overnight.

My hardhat had either blown off or fell off when I ducked and ran from the oncoming fireball. Several days later I saw my hardhat. The top of it was almost entirely blackened. I was told that my safety glasses were also blackened and the side shields were shattered. My right shoe, originally brown, was black with a burned area on the toe. I did not catch on fire. There was no electrical shock. My facial hair, the hair on my head below where the hardhat sat, and the hair on my hands and arms were singed. My throat felt scratchy, my voice was a little hoarse, and I felt heaviness in my lungs. I had a laceration on my right arm from the breaker door hitting it when it blew open. I had first and second degree burns on my face, ears, neck, arms, and hands.

If I had not been wearing my safety glasses, I probably would now have permanent eye damage, possibly blindness, in at least my right eye if not both. My ear plugs prevented burns further into my ear canals and muffled the bang I heard. From the appearance of my hardhat, my hair probably would have caught fire had I not been wearing it. If I hadn't been standing to the side of the breaker, while operating the open/close lever, my injuries would probably have been much worse, including third degree burns, possibly disfigurement and loss of body parts...possibly death.

I was not wearing a flash suit because I was operating the contacts of a 5kv motor starter with the door closed. According to the electrical safety policy in effect at the time, a flash suit was not required. That policy did require a long sleeve cotton shirt and leather gloves, but I had not been taught that nor had I seen it practiced at the plant in over four years working there. I had not seen it practiced at any other plant where I had worked. The short-sleeved shirt I was wearing was company issued for wearing on the job. The policy also stated that leather gloves were not required if the hand was on a non-conductive part. The open/close lever was plastic.

For over a month after the accident I "lived" in a recliner with pillows used to elevate my arms. Because of the burns to my face and ears, and the need to elevate my arms, I could not sleep in a bed. I got about four hours of sleep on a good night and usually a short nap or two during the day. My arms and left hand were covered with an artificial skin for about three weeks to aid the healing process. The bandaging of my arms and hands had to be replaced daily. I had weekly follow-ups at the burn unit which was thirty miles from home. I could not drive so a family member drove me. After about five weeks, the worst of the burns had healed over and I was cleared to return to work.

Instead of being allowed to go back to work, I was called in to my workplace for a disciplinary meeting the day before my released–to-work date and was given two weeks (84 hours) of unpaid leave and a "Final Written Warning" for not wearing a flash suit which could have prevented my injuries.

THAT'S NOT EXACTLY WHAT I MEANT BY "FLASH SUIT"!

One of the plant management types had called me the day following the accident to see how I was doing. During that conversation, he told me that the arc-flash "could have happened to anyone of us" and that I "hadn't done anything that hadn't been done a thousand times before". According to him, at that time, I had done nothing wrong.

Yet, now the injuries were my fault according to the company and I received the first disciplinary action of my life after coming close to losing that life due to faulty equipment!! (I think this is where I say, "Maaaaaa!")

According to coworkers involved in the investigation and the repair of the breaker, the arc-flash occurred because the motor was running at the time I pulled the handle and opened the breaker. They told me that the motor protection relay was wired wrong and caused the motor starter to restart the motor when power was restored to the 5kv bus. Also, a safety interlock mechanism that would have locked the handle and prevented the breaker from being opened was missing. Apparently the motor had been running for about a minute even though local and remote indicators showed it to be off and tripped. Just seconds before the accident, through radio communication with the on-shift supervisor in the control room, I was told that the motor would not start and that the control board showed it needed to be reset.

Three sources said the motor was off, two of the three said the breaker needed to be reset before the motor could be started, yet it was on.

You can never be too safe! There may be unknown hazards in the workplace. That motor starter that arc-flashed on me had been reset scores of times without incident. I had done so many times including the morning before the accident. It could have happened to anyone of us at the plant.

When my accident occurred, I was just six days shy of 29 years working in industrial plants without a lost-time accident. That record was gone in a flash.

About the Author

or

Who Is This Clown?

As I write this book, I am a 58 year-old Operations Technician II (referred to as a "B" operator in some plants) working for an independent power producer at a 425MW cogeneration power plant in southeast Texas. Born and raised in Long Beach, California, I moved from that area at age 36 and spent the next 17 years near Sacramento, California. In 2006, I transferred and moved to Texas.

My adult working career began with seven years in retail banking starting as bookkeeping clerk and working through the various positions to operations officer. Following that I worked three years in truck leasing starting as rental agent and moving up to district manager. I finally started making a livable wage in 1982 when I went to work for a major oil company as an operator trainee at an oil refinery. In 1989, I left the refinery to work at a cogeneration power plant and remain in that line of work today. My industrial plant experience has included two years as shift supervisor, four years as operations supervisor, and two years as the operations manager of three facilities.

I'm currently an operator technician by my own choice and not because of any demotion. Receiving three times more annual bonus and 25% more regular pay than the real workers made me feel guilty. I resigned from the operations manager job, spent some quality time with my kids, and then hired back in as a "B" operator. That's my preference at this stage of my life.

With a total of fourteen years of management experience in three different industries, I've spent many years on both the "us" and "them" sides of the workforce and have thus formed my general opinions of management types and the real workers, as I like to call them.

I'm not an expert on safety and I'm not a safety professional. I'm just a power plant operator who would like to see safety taken more seriously than what I've witnessed at some workplaces.

"Our lives begin to end the day we become silent about things that matter." – Martin Luther King, Jr.